Cover illustration: A Canberra TT.18 of No. 100 Squadron showing the dramatic effect of the black and yellow 'don't shoot me – I'm on your side' stripes carried by these target-towing variants (Squadron records via Fg. Off. Murphey)

1. No. 13 Squadron provided reconnaissance during the Suez Crisis with its PR.7s, supplemented by aircraft and crews from the UK squadrons. PR.7 WH799 became the only Canberra casualty of the war – and the RAF's only operational Canberra loss – when it was shot down by a Syrian Meteor on 6 November 1956.

CANBERRA AND B-57

KEN DELVE

Arms & Armour Press
London New York Sydney

Introduction

First published in Great Britain in 1988 by Arms & Armour Press Ltd., Artillery House, Artillery Row, London SW1P 1RT.

Distributed in the USA by Sterling Publishing Co. Inc., 2 Park Avenue, New York, NY 10016.

Distributed in Australia by Capricorn Link (Australia) Pty. Ltd., P.O. Box 665, Lane Cove, New South Wales 2066.

© Arms & Armour Press Ltd., 1988.
All rights reserved. No part of this book may be reproduced or transmitted in any form or by any means, electronic or mechanical, including photocopying, recording or via an information storage and retrieval system, without permission in writing from the publisher.

British Library Cataloguing in Publication data:
Delve, Ken
Canberra and B-57. – (Warbirds illustrated V. 51).
1. English Electric Canberra aeroplanes &
Martin B-57 aeroplanes
I. Title   II. Series
623.74'63

ISBN 0-85368-902-4

Edited by Roger Chesneau.
Typeset by Typesetters (Birmingham) Ltd.
Printed and bound in Great Britain by
The Bath Press, Avon.

On Friday 13 May 1949 VN799 took off from Warton Aerodrome for the first flight of the English Electric A.1 prototype. Despite what some might have considered to have been an inauspicious date, this was the starting point for one of the most successful British aircraft of all time – an aircraft that in 1989 will celebrate its fortieth anniversary and will still be in service in large numbers with the RAF and with other air forces.

The Canberra was designed as a 'blind' bomber, but the promised radar system did not materialise in time and the aircraft entered service as a visual, medium-level bomber, with 'Gee-H' as its main blind-bombing aid. The era of Bomber Command Canberra operations was short-lived, however, the early 1950s being the heyday of the Canberra Wings before the introduction of the V-bombers. Nevertheless, Canberra bombers, in a variety of guises, remained in front-line service until the early 1970s. The photo-reconnaissance Canberra force had an even longer life, the last operational squadron, No. 39, disbanding in May 1982, and even then a small number of PR.9s were retained for use with No. 1 PRU.

As the bomber squadrons disbanded and 'redundant' B.2s became available, new variants appeared to fulfil specific 'facilities' roles such as target-towing, calibration and EW/ECM training. The airframe surplus also boosted overseas sales of the Canberra as refurbished aircraft were transferred to air forces that were looking to update their inventories. In RAF colours, the Canberra saw active service at Suez and in Malaya, while many foreign air forces have also used their aircraft operationally, the latest being the *Fuerza Aérea Argentina* in the 1982 Falklands conflict.

Finally, the aircraft has been one of that breed of adaptable airframes highly suited to the trials and experimental field, and a great many have been used by military and civilian establishments in this role. One of the Canberra's greatest achievements was its acceptance for licence-build by the Americans as the B-57. The US Air Force saw the Canberra as a stop-gap until other aircraft entered service, but it was used extensively in the Vietnam War. The ultimate variant, the B-57G, was an ultra 'high-tech' aircraft equipped with advanced sensors and 'smart' weapons.

There is no way in which a short introduction can do justice to the Canberra story – and it has also been a problem deciding which photographs to leave out! I hope that this book gives a flavour of the widespread use and importance of the Canberra – one of the greatest aircraft of all time!

This book could not have been compiled without the help and advice of my Canberra colleague Peter Green. Individual photograph credits are given with each caption, uncredited illustrations being from the Ken Delve/P.H.T. Green Canberra Collection. The author would welcome correspondence from anyone with Canberra connections.

Ken Delve

◄2
2. No. 82 Squadron PR.7 WJ825 – a sleek, silver aircraft looking for a hole in the cloud through which to take a few photographs! (Squadron records)

▲3   ▼4

5▲

3. Canberra prototype VN799 being prepared for its first flight at Warton, 13 May 1949. Note the rounded fin and rudder, which was squared off after the first few flight trials during investigations into the aircraft's lateral stability. (BAe)

4. A truly aerodynamic shape: the neat circular fuselage so beloved of the Canberra's designer, W. Petter, and straight wings. The original design concept showed the engines mounted in the wing roots.

5. VN799 getting airborne for a display; note the Canberra nameplate on nose. The Canberra was designed as a 'blind' bomber, hence the solid nose for a radar system. It soon became obvious that the planned system would not be ready in time and so the aircraft was redesigned with a glazed nose position for visual bomb-aiming.

6. B.2 prototype VX165 on its way to a Farnborough display. The B.2 was given a glazed nose with a flat optical panel for the bomb sight, and crew composition was increased to three to include a bomb-aimer. VX165 wears standard Bomber Command camouflage.

6▼

▲7 ▼8

9 ▲

7. WH853, the first Shorts-built B.2. The outbreak of the Korean War highlighted the lack of suitable modern aircraft on the Western inventory, and in an effort to speed up the delivery rate of Canberras the decision was taken to start production at other aviation works. A. V. Roe, Handley Page and Shorts each undertook B.2 production.

8. No. 101 Squadron at Binbrook gave up its Lincoln bombers to become the first RAF jet bomber squadron. The first Canberra, WD936, was flown into Binbrook on 25 May 1951, and by the end of the year the full complement of eight had arrived. Here Canberra test pilot Roland Beamont (left) is seen with the Station Commander, Gp. Capt. Sheen.

9. A No. 101 Squadron formation over Lincolnshire in December 1951. The B.2 was powered by two Avon RA.1s each developing 6,500lb static thrust, giving the aircraft a top speed of over 450mph and a ceiling of 45,000ft. In essence this meant that the latest RAF bomber outperformed all current RAF fighters, a point aptly demonstrated during many Air Defence exercises. (Squadron records)

10. B.2 WD997 of No. 9 Squadron, September 1953. The third of the Binbrook squadrons, No. 9 formed in May 1952; note the Binbrook Wing flash on the nose. The original plan for Canberra deployment was for 24 squadrons each with ten aircraft, to be organized in Wings of four squadrons each. (MAP)

10 ▼

▲11

▲12 ▼13

11. WD965, a B.2 of No. 10 Squadron, photographed in 1957. The squadron re-formed with B.2s at Scampton in January 1953 and saw operational service with their aircraft during the 1956 Suez Crisis. (R. C. Ashworth)

12. WJ605 of No. 18 Squadron; note the squadron markings on the tip tank and the Honington Wing flash on the nose. WJ605 was lost in April 1962 when, serving with No. 45 Squadron, the aircraft broke up over the China Rock range. (E. Watts)

13. No. 21 Squadron B.2s over Aden. This unit operated B.2s from September 1953 until June 1957 and then B.6s until January 1959. The aircraft nearest the camera, WJ609, went back for conversion to a U.10 but instead was sold to the British Aircraft Corporation and became B.62 B-106 for the Argentine Air Force in 1971.

14. In the early 1950s a number of Bomber Command squadrons were involved with 'flag waving' tours in various parts of the world, the earliest being that by No. 12 Squadron to South America in 1952. Here No. 27 Squadron aircraft and crews prepare for their European tour of France, Italy, Greece, Turkey, Yugoslavia and Portugal in June 1954. (Squadron records)

15. WJ682 of No. 35 Squadron at Llanbedr, November 1961; note the squadron badge on the lower part of the fin. This aircraft was still in service with No. 100 Squadron in 1987. (A. Pearcy)

▲ 16

16. WJ641, a No. 50 Squadron B.2, in 1958. Bomb doors open, the aircraft sits on dispersal at Upwood where the squadron had been based since December 1955. The maximum bomb load of the B.2 was six 1,000-pounders, which was considered the war load. The majority of training sorties utilized 25lb practice bombs, a normal load being eight or twelve of these weapons, which simulate the ballistics of their larger brethren. (MAP)

17. WH720 of No. 57 Squadron, at Honington in 1955. One of the major problems of the Canberra was that the mid-wing position of the engines created asymmetric thrust in the event of an engine failure. (No. 57 Sqn.)

18. No. 100 Squadron was unique amongst the units of Bomber Command in that it acted as a trials unit for the Bomber Command Development Unit (BCDU). (Squadron records)

▲19
19. No. 61 Squadron B.2 WJ752 at Luqa, June 1957, with the accent on the canopy cover! During early trials the test pilot, Roland Beamont, had commented on the 'greenhouse' effect of the canopy, and shades such as the one here represented an attempt to keep cockpit temperatures down to a reasonable level. (M. Freestone)
20. WD986, a No. 100 Squadron B.12, in 1956; based at Wittering, the squadron operated B.2s and B.6s from March 1954 to September 1959. Entry into the Canberra was through the side door, shown here, then up to the seat for the pilot and a crawl to the nav compartment behind the pilot for the navigator.
21. WD944 of No. 101 Squadron, near Binbrook, December 1951. The dotted area behind the canopy indicates the hatch over the nav compartment. (RAF Binbrook)
▼20

22. B.2 WK146 of No. 102 Squadron, Blackbushe, 1956. With all suitable airfields in the UK full, the decision was taken to put the last of the bomber Wings into RAF Germany, but as part of Bomber Command. No. 551 Wing, comprising Nos. 102, 103, 104 and 149 Squadrons, was based at Gütersloh from October 1954 to August 1956. (E. Watts)
23. B.6 WJ768 of No. 109 Squadron at Blackbushe, September 1956. The aerial visible on the lower part of the fuselage was for the Blue Shadow sideways-looking radar, an attempt to give the Canberra a 'blind' operating capability. No. 109 was the first squadron to have its aircraft fitted with this system and the unit undertook a series of trials to determine its effectiveness. Initial results were good and the system was adopted – but with mixed feelings and results. (E. Watts)

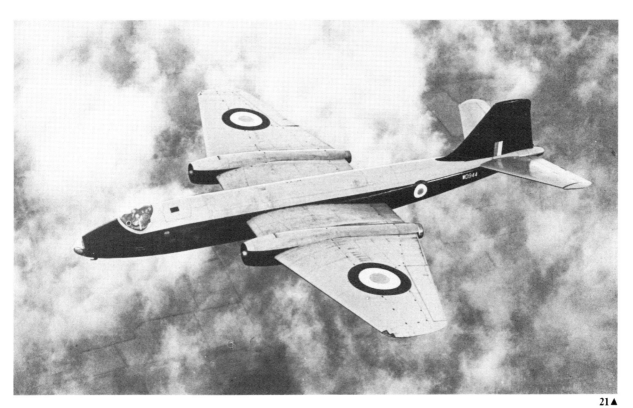

21▲

22▲ 23▼

▲24 ▼25

24. Canberra B.6 WJ773 of No. 139 Squadron. This unit, with No. 109, made up the Bomber Command Canberra Marker Force, operating initially from Hemswell but moving in January 1956 to Binbrook. In the marking role the Canberra dropped a mix of target indicators and flares. (C. Waterfall)

25. No. 139 Squadron B.2 WJ609 at Llanbedr in September 1959, on trials with the RAE and still carrying No. 139's markings despite having left the squadron four years earlier.

26. B.2 WJ616 of No. 199 Squadron was one of only two Canberras operated by this specialist ECM unit. (MAP)

27. With large numbers of Canberras entering service, an Operational Conversion Unit became essential, and so No. 231 OCU at Bassingbourn became the home for all Canberra training in February 1952. At one stage the unit had some 80 Canberras on strength in its two bomber squadrons and one recce squadron, operating B.2s, T.4s and PR.3s on a variety of conversion courses. This is B.2 WH641/'A'. (MAP)

28. Four B.2s from the OCU, photographed in 1953. The OCU became well known, and admired, for its close-formation displays by four aircraft. It was a spectacular sight to see such large aircraft flown in a tight box and on an impressive bomb-burst finale.

▲29   ▼30

29. T.4s of No. 231 OCU in a box, led by the prototype T.4, WN467. The T.4 had side-by-side seating for the pilots – a very cramped arrangement – with the navigator in his usual 'black hole' behind.

30. The recce squadron of the OCU operated a number of PR.3s; here WE139 is seen on a visit to Pershore. Note the plaque commemorating this aircraft's victory in the London to New Zealand Air Race, and also the OCU's leopard head on the fin.

31. The Canberra was seen as an ideal reconnaissance platform and so a PR variant of the B.2 entered service in late 1952. The aircraft was equipped with long focal length cameras for the strategic recce (SR) role of pre- and post-strike photography. The main recce element of Bomber Command came to be based at RAF Wyton (opening that station's continuing association with the Canberra) with Nos. 58, 82 and 540 Squadrons, and within two years the squadrons were exchanging their aircraft for the more powerful and longer-range PR.7. The first unit to re-equip was No. 540, although No. 542 had been operating PR.7s from Wyton since the previous June. The photograph shows a PR.3 of No. 540 Squadron, which operated this mark of Canberra at Benson and then at Wyton. The first aircraft was taken on charge in December 1952.

▲32 ▼33

32. PR.7 WJ815 of No. 58 Squadron, one of only two Canberra units to operate all three recce variants – eventually receiving the PR.9 for a short period.
33. The B(I).8 prototype, VX185, which made its first flight on 23 July 1954. The B(I).8 was built to meet an Air Staff Requirement (ASR) for a low-level interdictor version of the Canberra to be equipped with a wide range of conventional weapons.
34. B(I).8 WT346 of No. 16 Squadron, summer 1972, as the unit exchanged its Canberras for Buccaneers. This aircraft had been SOC in 1971 but was retained at Laarbruch as a decoy. (B. Pickering)

34 ▲

▲ 35 ▼ 36

35. WT339 of No. 14 Squadron. No. 14 was the fifth of the Canberra B(I).8 squadrons in Germany, although it was in truth brought into being by the renumbering of No. 88 Squadron. This particular aircraft is now in use at Cranwell as a ground instructional (GI) airframe. (Squadron records)

36. B(I).8 WT329. The Mk.8 was a true pilot's aeroplane and was well liked by all who flew it. The bubble canopy gave excellent all-round visibility, an important consideration in the low-level tactical scenario; it was still a fixed canopy, however, entry being gained through the door on the starboard side. Note the additional side windows at the nose. (BAe)

37. B(I).8 XM276 of No. 3 Squadron. When the squadron disbanded in 1972 most of the aircraft went into store at St. Athan, whence a number were sold to Marshalls of Cambridge for their B(I).68 contract for the Peruvian Air Force. (Squadron records)

38. Low-level (note the standing figure), this time over the El Adem range in March 1969. The photograph shows up the belly bulge of the gun pack. El Adem was the venue of many a weapons detachment, and flying at ultra low level over the desert became a favourite pastime.

▲39 ▼40
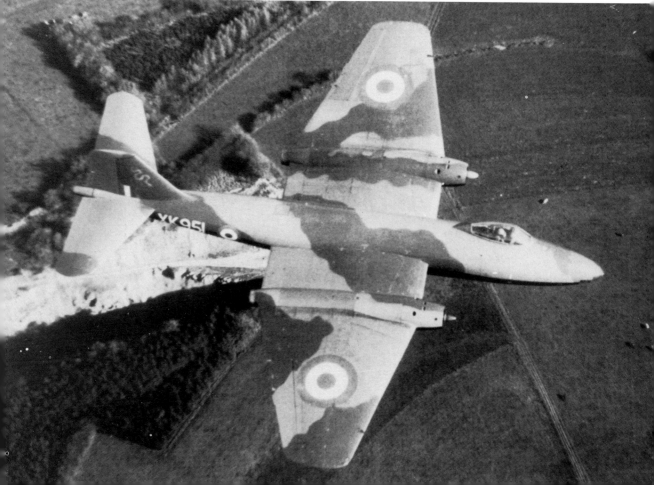

39. One of the all-time-great photographs: a No. 213 Squadron B(I).6 during a low pass, by Eric Tilsley. The B(I).6 was the interim interdictor – a Mk.6 with a gun pack and underwing pylons. However, No. 213 kept its B(I).6s until 1969, and held its own against the (theoretically) superior B(I).8s! The hardpoints under the wings carried bombs or rocket pods.
40. The B(I).8 packed a mighty punch, its gun pack containing four 20mm Hispano cannon, with 525 rounds per gun. Here a No. 88 Squadron aircraft, XK951, is seen over Germany in April 1960.
41. PR.7 WH775 of No. 31 Squadron. The RAFG Canberra recce establishment comprised four squadrons, No. 31 being the second of these, forming at Laarbruch in March 1955 with PR.7s. All were primarily equipped for the medium/high-level PR role of pre- and post-strike reconnaissance, although low-level tactical PR was introduced later.
42. PR.7 WH803 of No. 17 Squadron, October 1969. A few months later the squadron said farewell to its Canberras; it had been the last RAFG Canberra recce unit to form, having received its first PR.7s at Wahn in June 1956.
43. PR.7 WT517 of No. 80 Squadron – the fourth of the RAF Germany reconnaissance squadrons.

▲44 ▼45

44. The Suez Crisis of 1956 saw the major operational employment of the Canberra force during its years with Bomber Command. Operation 'Musketeer' called for the Canberras and Valiants to neutralize the Egyptian Air Force by destroying its airfields. Eventually seven squadrons (Nos. 10, 15, 18, 27, 44, 62 and 139) operated out of Nicosia and four (Nos. 9, 12, 101 and 109) out of Hal Far, Malta. Between 1 and 5 November the Canberras flew a total of 278 operational sorties, although this does not include the numerous PR sorties flown by the reinforced No. 13 Squadron. This is No. 61 Squadron B.2 WH907 out of Nicosia, photographed in December 1956 after the Crisis. This squadron trained at short notice in the SDB (shallow dive-bomb) delivery of TIs but in the event only flew 'ops' in the bombing role. WH907 was eventually struck off charge at at No. 19 Maintenance Unit in November 1973. (M. Freestone)

45. No. 12 Squadron B.6s at Hal Far, October 1956. The longer-range B.6s were based in Malta but they had to reduce their bomb load to four 1,000lb bombs because of the distance to and from the targets. (B. A. Crook)

46. A No. 27 Squadron line-up at Nicosia during the Suez Crisis.

▲ 47

▲ 48 ▼ 49

47. No. 69 Squadron PR.3s at Laarbruch; the only one of the RAFG units to operate the PR.3, No. 69 moved to Malta in 1958. The second aircraft in line, WE139, carries yet another version of the 'Winner' plaque.

48. Under the terms of the Baghdad Pact, Britain agreed to station strike aircraft in Cyprus, and this led to the formation of the Akrotiri Strike Wing (ASW) of four squadrons. Initially all were equipped with B.2s, but in the early 1960s the Wing received B.15s and B.16s to enable it to field a wider ranger of conventional weapons, including underwing rocket pods and, for two of the squadrons, AS.30 air-to-surface missiles. Shown is No. 73 Squadron B.2 WK111, October 1960.

49. B.2 WH655 of No. 249 Squadron, the last of the squadrons to arrive at Akrotiri (in October 1957). Like the other units, No. 249 carried markings on fin and tip tanks and, again like its sister-squadrons, changed these at regular intervals. The markings shown here are perhaps the most 'subtle'! (R. C. Ashworth)

50. Canberra B.16s of No. 6 Squadron out of Akrotiri. Only minor equipment differences singled out the B.16 from the B.15 which served with the other Akrotiri squadrons – although in practice, especially when the ASW became a single unit, aircraft were exchanged amongst the units.

51. B.16 WJ776 of No. 6 Squadron, which received this model from February 1962. By the mid-1960s all B.15s and B.16s were grouped together as the Akrotiri Strike Wing, an unpopular move with air crew and ground crew alike.

52. WH968 of No. 32 Squadron with an RP pod on the underwing pylon. (P. Tomlin)

50▲

51▲   52▼

▲53

▲54

53. WH964 of No. 249 Squadron showing yet another variation of aircraft marking: here the tip tank is painted with a gold arrow.
54. B.15 WH967 carrying two Nord AS.30 missiles. The AS.30 was the last of the weapons introduced to the Canberra and was used by the ASW units and No. 45 Squadron in the Far East Air Force (FEAF). (FGS)
55. B.16 WT374 of the Akrotiri Strike Wing. The move to ASW markings of a 'flamingo above crested waves over a lightning flash' was made in 1965 and was followed by the introduction of centralized servicing whereby aircraft lost all trace of squadron identity.
56. B.16 WJ778 of the ASW. One way in which the squadrons got around the 'centralization policy' was to add their own unit marking to the ASW crest – in this case that of No. 249 Squadron.
57. AOC Aden's Canberra WH727, 'Queen of the Arabian Skies'. This white B.2 was used by the AOC to visit the various extremes of his far-flung Command.

55 ▲

56 ▲ 57 ▼

▲58

▲59 ▼60

58. No. 39 Squadron re-equipped with the ultimate Canberra, the PR.9, in late 1962. Here XH165 sits in the sun in its original silver finish and sporting tip tanks. The PR.9 was the only Canberra to have a conventional opening canopy.
59. PR.9 XH131 of No. 39 Squadron in a camouflaged paint scheme as introduced in the late 1960s. Pilot access was via the removable ladder, whilst the navigator climbed in through the hinged nose.
60. The 6 ATAF area, the southern flank of NATO, was devoid of reconnaissance resources and so No. 69 Squadron was redeployed from RAFG to Malta in 1958. On arrival the unit was renumbered No. 39 Squadron, opening a 25-year Canberra PR period for that unit starting, as here, with the PR.3.

61. No. 13 had been the first Canberra squadron in the Mediterranean theatre, being first stationed at Akrotiri with its PR.7s during the Suez Crisis. Like all PR units, it ranged far and wide, often well outside its 'home skies'. This No. 13 Squadron PR.9, XH135, was photographed by a No. 39 Squadron aircraft.
62. No. 101 Squadron B.2s over Malaya in 1955, during Operation 'Mileage', the Bomber Command reinforcement for anti-terrorist 'ops' during the Malayan Emergency. All the squadrons of the Binbrook Wing took their turn, but No. 101 was unique in having two such deployments. The Canberras were employed not only to bomb known supply/HQ areas but also to 'area-bomb', in order to drive the terrorists into pre-planned ground ambush zones.

▲ 63

▲ 64 ▼ 65

66 ▲

63. The FEAF squadrons at Tengah in the early 1960s: No. 45 Squadron B.15s are in the foreground, with No. 81's PR.7s lined up behind. No. 45 became the resident FEAF Canberra bomber squadron in December 1957 with B.2s, re-equipping with B.15s in 1962.
64. WH646 of No. 45 Squadron. The squadron's B.2s saw combat service in the late 1950s and early 1960s during Operation 'Firedog', the Malayan Emergency.
65. WH874 jacked up ready to receive 500lb practice bombs, Tengah, 1960. Note the use of a wheeled sun-shade rather than the clip-on variety. (R. Evans)
66. 1,000lb bombs are loaded into a No. 45 Squadron B.2. Despite the fact that the aircraft had to be jacked up to give access to the bomb bay for the trolley, turn-round times for six 1,000lb bombs were a matter of minutes – a testimony to the skill of the armourers. (R. Evans)
67. No. 81 Squadron PR.7s start up at Tengah. The black plumes are starter cartridge fumes; when the starboard engine was started this black cloud often went into the cockpit – an unforgettable smell and taste!

67 ▼

▲68 ▼69

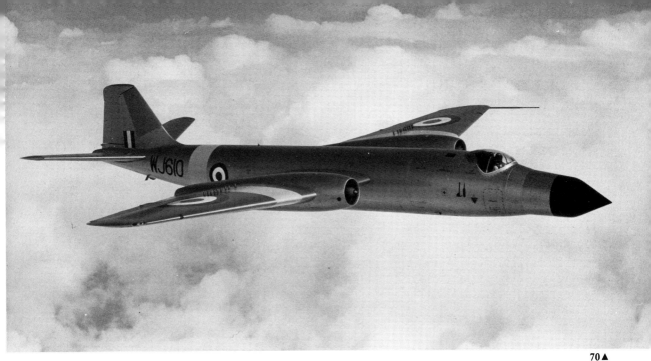

68. No. 81 Squadron PR.7s were FEAF's main recce element from 1960 to 1970, although during hectic periods detachments from the UK and Mediterranean squadrons were called upon.
69. In the 1960s the Canberra began a second lease of life, refurbished and modified aircraft appearing in a variety of new designations for 'facilities' squadrons. One such unit was No. 85 Squadron, which became a target facilities unit in March 1963 with T.11s and T.4s, later receiving B.2s and T.19s as well. This shows a No. 85 Squadron line-up at Binbrook, with B.2 WJ567 nearest the camera.
70. The T.11 was designed as an air intercept radar trainer for fighter crews and was given an Airpass radar in the nose. T.11s first served with No. 228 OCU at Leeming in 1958. The photograph shows T.11 WJ610.

▲ 71

71. WJ640 of No. 85 Squadron over the threshold, with flaps down and throttles closed.
72. A fine airborne study of B.2 WJ728, No. 100 Squadron, in 1973.
73. TT.18 WJ639, No. 7 Squadron, 1976. (MAP)
74. B.2 WH666 of No. 56 Squadron. This squadron ran a target facilities flight of three aircraft to support its Lightnings in Cyprus. On leaving No. 56's TFF, WH666 went to No. 100 Squadron at Marham and subsequently became one of the Marham aircraft to go to Zimbabwe (as 2250) in March 1981.
75. TT.18 WJ721 of No. 7 Squadron over Gribben Head, 1971. Each aircraft carried two Rushton target winches for either a Rushton Mk.2 target or a conventional sleeve target.

▼ 72

73▲

74▲ 75▼

76. In the early 1960s two squadrons, Nos. 97 and 98, received Canberras for calibration duties, although No. 97 also operated a number of specialist ECM/EW B.6s. This is a No. 98 Squadron B.2, WK162, 1970.
77. A 1970 photograph of B.6 WJ775 of No. 51 Squadron, the 'specialist' unit which operated a variety of aircraft types on calibration and EW duties until the early 1970s. The special equipment was stowed in retractable bins within the fuselage.
78. No. 360 Squadron T.17 WF916, Wyton, 1984. The T.17 was developed to fulfil the EW/ECM training role previously performed by other modified Canberras. The lumpy nose contains the various aerials of the 'wiggly amps' kit. (S. G. Richards).

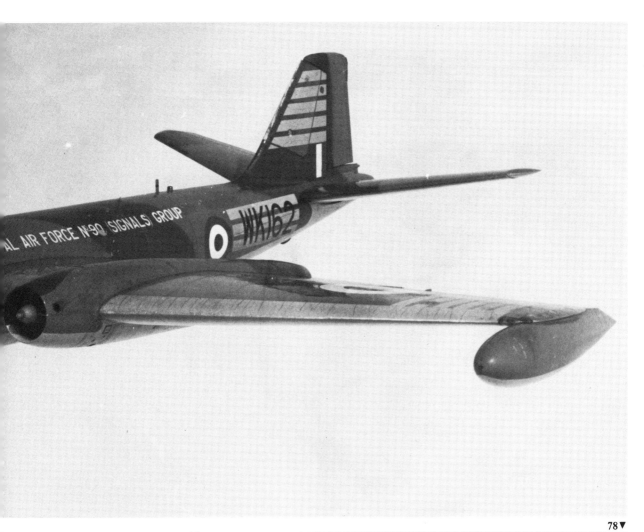

79. T.17A WJ607 of No. 360 Squadron, Wyton, July 1986. The hemp colour scheme and a change to the avionics, especially the ECM kit, makes this a T.17A rather than a straight T.17. (J. B. Hale)

80. PR.7 WJ821 of No. 58 Squadron. The squadron's PR.7s shouldered the UK reconnaissance burden until 1970, when the unit was disbanded. The two square windows just forward of the engine are the camera flats for the large-format oblique camera and the starboard F95 cine camera.

81. PR.9 XH168 of No. 39 Squadron, 1971. The squadron returned to the UK in October 1970 and acquired all the remaining PR.9s from the other PR units. The bubble canopy, offset to the port side, gave the pilot excellent visibility.

81▼

▲ 82 ▼ 83

82. PR.7 WT537 of No. 13 Squadron, Wyton, 1979. The last RAF squadron to leave Malta, No. 13 returned to the UK in 1978 and operated alongside No. 39 in worldwide tasking until its disbandment in 1981.

83. XH166 of No. 39 Squadron, Wyton, 1981. The antenna on the fin is the forward sector aerial of the radar warning receiver fitted to some PR.9s from 1976 onwards.

84. PR.7 WH796 of No. 39 Squadron, at Wyton in November 1971. No. 39 operated two PR.7s in the early 1970s, partly to overcome the high task workload which the PR.9s were unable to meet on their own.

84 ▲

▲85

▲86 ▼87

85. B.2 WH919 of No. 231 OCU, Cottesmore, 1974; note the leopard head on the fin. The OCU shrank in size as the number of Canberra squadrons contracted and the requirement for new crews was reduced.
86. The *Fuerza Aérea Argentina* was one of the last overseas customers for the Canberra, buying six ex-RAF B.2s and two T.4s in 1970. These aircraft saw operational service during the 1982 Falklands War and two were shot down. This photograph shows B.62 B-110, which was shot down by an AIM-9L fired by a RN Sea Harrier on 1 May 1982. (P. A. Tomlin)
87. B.20 A84.236 of No. 2 Squadron, the first RAAF unit to give up its Lincolns for Canberras, at Amberley. In April 1967 the squadron deployed to Phan Rang in Vietnam for what proved to be a four-year operational stint, during which time it established a reputation for pin-point weapons delivery.
88. Chile, the only overseas user of the PR.9, received three ex-No. 39 Squadron aircraft in late 1982. One of these has since crashed in a flying accident.
89. India has been the biggest customer for the Canberra, with over 100 aircraft acquired in bomber, trainer and reconnaissance variants. An initial order for 80 aircraft was placed in 1957, and the aircraft saw operational service in the Congo in 1960 as part of the UN Force and in the 1965 and 1971 wars with Pakistan. This is B(I). 58 IF 919, of No. 35 Squadron, over the Persian Gulf in the late 1960s. The B(I).58 was almost identical to the RAF B(I).8.

▲ 90

90. PR.57 P1099 awaits delivery to the IAF. A PR.7 equivalent, the PR.57 had minor changes of equipment to suit the IAF. The aircraft are still (1988) operational with No. 106 Squadron.
91. B.2 WJ605 of No. 75 Squadron RNZAF, late 1950s. This squadron's aircraft were on loan from the RAF as part of the Commonwealth Strategic Reserve until 1962, when they were handed back. WJ605 went to No. 45 Squadron RAF at Tengah.
92. NZ6103, a B(I).12 of No. 14 Squadron RNZAF. No. 14 re-equipped in October 1959 and became operational on 1 May the following year. During 'Confrontation' in the early 1960s the unit moved to forward bases in the FEAF area. (RNZAF)

93. Ecuador ordered six B.6s in May 1954 for counter-insurgency (COIN) duties with *Esc. de Bombardeo 2123* at Quito.
94. France ordered six B.6s in 1954 for use as trials and development platforms, primarily for systems and weapons. Most of the aircraft served with the trials unit at Cazoux. This is F316, with radar nose and a missile on its pylon.
95. (Overleaf) Peru was another of the COIN users, placing a number of small orders for Canberras to equip two squadrons of *Grupo do Bombardeo 21*. The first aircraft was delivered in May 1956. Here 244 displays a range of underwing RPs and ASMs. (BAe)

▼ 91

92 ▲

93 ▲ 94 ▼

▲ 96

▲ 97 ▼ 98

96. The Rhodesian Air Force was another operator whose Canberras saw active service, in this case on anti-terrorist work. The rugged nature of the Canberra and its heavy and diverse weapon load made it very suitable for this role. B.2 2005 (ex-RAF WH867) is seen here at Durban in April 1970.
97. B(I).12 455 of No. 12 Squadron, South African Air Force, at Waterkloof, February 1980. The SAAF purchased six B(I).12s and three T.4s in 1962 and these aircraft have been frequently employed on counter-terrorist operations.
98. B.82 1233 of the *Fuerza Aérea Venezuela*, June 1979. Many Canberra B.82s are still operational and, with a wide range of weapons (including ASMs) available, continue to form a potent element within the *FAV*. (MAP)
99. T.52 52002 (ex-RAF B.2 WH905) at Malmen Museum, 1975. Sweden bought two ex-RAF B.2s for use as elint aircraft in the 1960s. Both were operated by the F8 Wing at Barkby and were retired in 1973. (Swedish Air Museum)
100. B(I).8 0923 destined for the *FAV*, photographed at Warton in 1966. Venezuela was the first overseas customer for the Canberra, the initial order being placed in 1953; subsequent orders increased the total to over thirty aircraft of a variety of marks. (BAe)

▲101

▲102 ▼103

101. In 1966 West Germany bought three B.2s for trials and experimental duties, and since then they have performed a wide range of roles. Two aircraft are still in use, one of their primary tasks being aerial survey. This is B.2 99+34 (ex-RAF WK137), in 1979.

102. A number of Royal Navy units have operated Canberras in support roles since the late 1950s. WH876 was a D.14 of 728B Naval Air Squadron, seen here over Hal Far, Malta, in 1961. This aircraft, together with a number of other converted B.2s, was used as an unmanned target for guided weapons trials.

103. B.2 WK147 of 776 Fleet Requirements Unit, Hurn, 1969, operated by Airwork Services Ltd. as a target facilities unit.

104. B.2 WD952 in use with Bristols as the Olympus test-bed.

105. The second Canberra prototype, VN813, was used by De Havilland Engines as a test-bed for the Spectre rocket motor, seen here during a test firing. (DH Engine Co.)

106. WJ643, a modified B.2, in use with the Ferranti Flying Unit (FFU) at Turnhouse, Edinburgh. The FFU operated some eight Canberras over a period of time on systems and avionics trials. (B. Pickering)

104 ▲

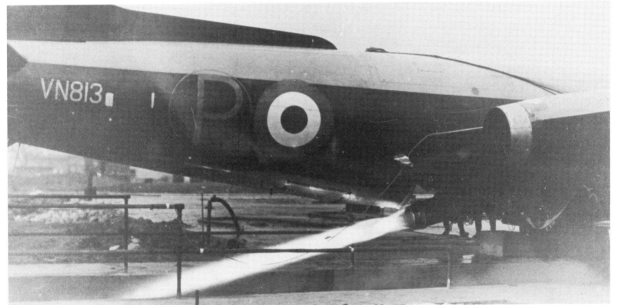

105 ▲ 106 ▼

▲107 ▼108

109▲

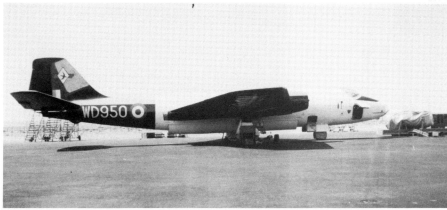

110▲

107. WJ644 launching a Firestreak air-to-air missile over in the Aberporth range – one of a number of missile trials involving the Canberra, which was an ideal platform for such duties because of its inherent stability.
108. A B.2 operated by Flight Refuelling Ltd. on air-to-air refuelling trials. Note the Hose Drum Unit (HDU) in the bomb bay. (Flight Refuelling Ltd.)
109. A&AEE hybrid WV787, the 'icing tanker'. Equipped with a spray-bar system below the rear fuselage, this aircraft was used to spray other aircraft (in this case another Canberra) with water droplets to test icing conditions and ice-prevention systems.
110. Swifter Flight B.2 WD950 at El Adem. This flight of six aircraft was used to investigate the effects of high-speed, low-level flight on aircraft and crews. (R. A. Walker)
111. B.2 WH876, used by the A&AEE for ejection seat trials.

111▼

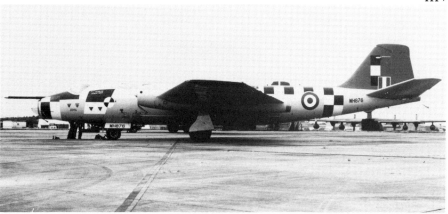

▲112 ▼113

112. A trio of RAE aircraft used for radar development, in the case of WJ646 for the AI.18 programme. (R. Henry)

113. The Canberra was one of that very select breed of aircraft taken up by the Americans for licence-build. Following detailed studies in the early 1950s, WD932 went to Andrews AFB in February 1951 for a 'fly-off' competition. A 'hands-down' winner, the Canberra was adopted for production by Martin as the B-57 Canberra, two aircraft, WD392 and WD940, being purchased as patterns. The B-57A (shown) was almost identical to its parent Canberra B.2 and made its first flight in July 1953.

114. Only eight aircraft came off the production as B-57As, the remainder being modified during production into RB-57A reconnaissance variants, which went on to equip the first unit, the 363rd TRW, in 1954. As RB-57As were retired from operational squadrons they were often handed to the Air National Guard; 21426, for example, saw service with the Michigan ANG, June 1968.

115. Even as the B-57A was going into production, major changes to the aircraft layout were underway, leading to the B-57B with its excellent tandem canopy. The B-57B became the mainstay of the B-57 units deployed in Europe and the Far East. (Martin)

▲116 ▼117

116. A B-57B of the 13th Bomber Squadron; the nose inscription reads 'The 1st B-57 to land in Korea'. Aircraft of the 3rd Bomb Group were based in Japan but rotated to Korea and instead of the standard glossy black finish were left in natural metal. (R. Evans)
117. B-57B 33862 of the 822nd BS, 38th BG, which was based at Laon, France, in the late 1950s. (Martin via B. Robertson)
118. The B-57s were given a second lease of life with the outbreak of the Vietnam War. The first B-57s went to Bien Hoa in August 1964 but were restricted to unarmed recce flights until the following year. The photograph shows a B-57B of the 8th TBS, in 1976.
119. RB-57C 33842 of the 55th WRS (Weather Reconnaissance Squadron). The B-57C was a trainer version of the B-57B, identical except for a second set of controls in the rear cockpit.

▲120
120. B-57E 54258, seen here in 1970. The B-57E was built as a target-tug and given a modified tail section equipped with two target canisters.
121. EB-57E 54253 of the 4713rd Defense Systems Evaluation Squadron (DSES) at Stewart AFB, May 1964. A number of B-57Es were diverted to EW/ECM training roles under this new designation.
122. Designed for the high-level reconnaissance and air-sampling task, the RB-57F was given 122ft-span wings and more powerful engines. The aircraft was never a total success because of structural problems. This RB-57F was assigned to the 58th WRS, 1970.
123. The ANG operated B-57s from 1958 to 1972 for a variety of support tasks. This is EB-57B 1499, of the 117th DSES, Kansas ANG, at Offutt AFB in May 1975.
124. EB-57B 21505 of the 134th DSES, Vermont ANG – 'The Green Mountain Boys'.

▼121

122▲

123▲ 124▼

▲125
125. The ultimate variant of the B-57 was the specialized B-57G, optimized for the night interdictor mission and equipped with a wide range of sensors and 'smart' weapons. Aircraft were modified to this configuration and sent back to Vietnam, where they proved highly successful, prowling the night skies and hitting ground targets with exceptional accuracy.

CANBERRA DATA

The Canberra has appeared in no fewer than sixteen different marks of UK production for RAF and Royal Navy use, with an even larger number of designations being used for the export versions of what are essentially the same variants. However, the list can be reduced to just three basic airframes: the B.2 and its variants; the B.6 and its variants; and the unique PR.9.

	B.2	B.6	PR.9
ENGINE			
Type	2 × Avon RA3 (Mk. 101)	2 × Avon RA7 (Mk. 109)	2 × Avon Mk. 206
Thrust	6,500lb s.t.	7,400lb s.t.	11,250lb s.t.
Starter	Single cartridge	Triple cartridge	Avpin
DIMENSIONS			
Length	66ft 8in	66ft 8in	66ft 8in
Span	65ft 6in	65ft 6in	67ft 10in
Height (fin)	15ft 7in	15ft 7in	15ft 7in
WEIGHTS			
Max. take-off	42,000lb	55,000lb	57,500lb
Max. landing	31,500lb	40,000lb	40,000lb
FLIGHT LIMITS			
Max. speed	450kt	450kt	450kt
Max. with tip tanks	365kt	365kt	365kt
Ceiling	Over 45,000ft	Over 45,000ft	Over 50,000ft

WEAPONS

It is not possible to give a standard weapons break-down for the Canberra as different marks and sub-marks carried different combinations of weapons. The 'basic' bomber versions of B.2 and B.6 were designed for a maximum bomb load of 6,000lb with the option of 500lb, 1,000lb or even 4,000lb bombs. Interdictor variants were equipped with cannon packs, underwing RP pods and air-to-surface missiles and the PR variants carried a combination of cameras and flares. Some of the overseas users had an even wider range of weapons available. In essence, almost any weapon that has been available since 1945 has, at some time, been tried on a Canberra!

DESIGNATIONS*

B.2	Main RAF bomber variant.
PR.3	Reconnaissance version of the B.2 for RAF.
T.4	Dual-control trainer based on B.2.
B.6	RAF bomber.
B(I).6	Interim interdictor variant.
PR.7	Reconnaissance version of B.6.
B(I).8	Interdictor, based on B.6 airframe.
PR.9	Ultimate reconnaissance version.
U.10	B.2 conversion to unmanned target.
T.11	B.2 conversion to radar trainer.
B(I).12	B(I).8 for RNZAF.
T.13	T.4 for RNZAF.
D.14	B.2 conversion to unmanned target.
B.15/E.15	B.6 conversion to interdictor facilities aircraft.
B.16	B.6 conversion to interdictor.
T.17	B.2 conversion to ECM trainer.
TT.18	B.2 conversion to target-tug.
T.19	T.11 conversion to facilities aircraft.
B.20	Australian-produced B.2.
T.21	Australian-produced T.4.
T.22	PR.7 conversion to radar trainer.
B.52	Ethiopian B.2.
B.56	Peruvian B.6.
B(I).56	Peruvian interdictor.
PR.57	Indian PR.7.
B(I).58	Indian B(I).8.
B.62	Argentinian B.2.
T.64	Argentinian T.4.
B(I).66	Indian B.15.
B(I).68	Peruvian B(I).8.
B.72	Peruvian B.2.
T.74	Peruvian T.4.
B.82	Venezuelan B.2.
B(I).82	Venezuelan B.2 interdictor.
B(I).88	Venezuelan B(I).8.

*Not including prototypes or 'one-offs'